DISCARD

Invertebrates

Earthworms, Leeches and Sea Worms

Annelids

Beth Blaxland

for the Australian Museum

MACMILLAN
LIBRARY

First published in 2002 by
MACMILLAN EDUCATION PTY LTD
627 Chapel St, South Yarra 3141
Associated companies and representatives throughout the world.

National Library of Australia
Cataloguing-in-Publication data

Blaxland, Beth.
    Annelids.

    Includes index.
    For primary school students.
    ISBN 0 7329 8109 3.

    1. Annelida – Juvenile literature.
    I. Title. (Series: Invertebrates (South Yarra, Vic.)).

592.6

Edited by Anna Fern
Text design by Polar Design Pty Ltd
Cover design by Polar Design Pty Ltd
Illustrations by Peter Mather, Watershed Art and Design
Australian Museum Publishing Unit: Jenny Saunders and Kate Lowe
Australian Museum Series Editor: Deborah White
Australian Museum Scientific Adviser: Dr Pat Hutchings

Printed in China

Acknowledgements
The author and the publisher are grateful to the following for permission to
reproduce copyright material:

Cover photograph: Fire worm, courtesy of Karen Gowlett-Holmes/Nature Focus.

Alex Steffe/Lochman Transparencies, pp. 16, 18 (top); Australian Museum/Nature
Focus, pp. 26 (bottom), 29 (Bottom); Carl Bento/Nature Focus, pp. 4 (top), 14;
Clay Bryce/Lochman Transparencies, p. 4 (bottom); Dennis Sarson/Lochman
Transparencies, pp. 12, 21 (top), 27; Eva Boogaard/Lochman Transparencies,
pp. 3, 7 (bottom), 17 (bottom), 20; Jay Sarson/Lochman Transparencies, p. 28;
Jiri Lochman/Lochman Transparencies, p. 5; Karen Gowlett-Holmes/Nature Focus,
pp. 7 (top), 8, 9, (all), 11, 17 (top), 18 (bottom), 19, 21 (bottom), 22, 23 (all), 25, 29 (top);
Pavel German/Nature Focus, p. 6 (bottom); Peter & Margy Nicholas/Lochman
Transparencies; Peter Marsack/Lochman Transparencies, pp. 15, 24 (all), 30;
Philip Griffin/Nature Focus, p. 6 (top), 26 (top).

While every care has been taken to trace and acknowledge copyright the publisher
tenders their apologies for any accidental infringement where copyright has proved
untraceable. Where the attempt has been unsuccessful, the publisher welcomes
information that would redress the situation.

Contents

Glossary words

When a word is printed in **bold**, you can look up its meaning in the Glossary on page 31.

What are annelids?

Annelids are a group of invertebrate animals. An invertebrate is an animal that does not have a backbone. There are many different kinds of invertebrates. Some other examples of invertebrates are spiders, crabs and millipedes. Many other animals are invertebrates too. Can you think of any?

There are three main kinds of annelids. Some live on land, some live in the sea and some live in **freshwater** ponds, lakes and rivers. The main kinds of annelids are:

◎ earthworms

◎ leeches

◎ sea worms (which include bristle worms, beach worms and tube worms).

▲ A leech is an annelid.

Fascinating fact

There are more than 20 000 different types of annelids.

◄ A sea worm is an annelid. This sea worm is called a tube worm because it lives inside a tube. Most of its body is hidden inside the tube.

How do you say it?

annelids:	*an-a-lids*
invertebrate:	*in-**vert**-a-brate*
millipedes:	*mill-i-peeds*

General features of annelids

Some kinds of annelids look very different to others. For example, most sea worms look very different to earthworms. Scientists put these different-looking animals into the same group because they are all closely related. Scientists know these animals are closely related because their bodies all have the same general features.

Earthworms, leeches and sea worms are all annelids because they are all invertebrates that:

◎ do not have a hard skeleton

◎ have a soft body

◎ have a long body made of many similar parts called **segments**. These segments can look like little rings around the body, but sometimes they are hard to see. The segments are filled with liquid and this helps annelids keep their body shape and move about.

Ⓥ **An earthworm is an annelid. Annelids are worms that have a body made up of many segments. The lines around the earthworm's body show where one segment joins onto another.**

Annelid bodies

Annelids have a long, soft body with a small head at one end. Some annelids have a tiny head that is very hard to see. An annelid's body is made up of many segments. The segments often have special features on them such as paddles, bristles or suckers. Some annelids, such as earthworms, have similar-looking segments all along their bodies. Other annelids, such as tube worms, have different-looking segments on different parts of their bodies.

The bodies of earthworms, leeches and sea worms each look a little different.

Earthworms

Earthworms have a long body with a tiny head. The head is very simple and does not have eyes or **tentacles**. The body has many segments and there are tiny bristles under each segment that can stick out of the earthworm's skin. These bristles are so small that a **microscope** is needed to see them.

△ **This is an earthworm. Its body segments look very similar all along its body.**

Leeches

Many leeches have a long body that is narrow at the head end and wider at the tail end. The head does not have tentacles, but it can have up to 10 tiny eyes.

Leeches have suckers under their bodies that help them move around and feed. A big sucker is found under the tail end. Most leeches also have a smaller sucker under their head. This sucker surrounds the mouth.

▽ **This leech is on a person's leg.**

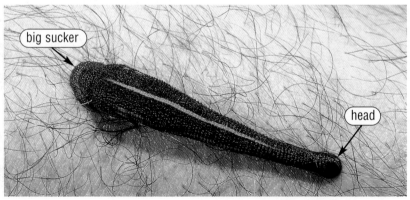

big sucker

head

Did you know

Earthworms were given their name because many of them live on land and burrow into the earth or ground. Some earthworms live in water. They are called earthworms too because they have the same body features as other earthworms.

Sea worms

A sea worm's body has many bristles. These often stick out from the ends of flat paddles that grow on the sides of their bodies. Some sea worms, such as beach worms, have two big paddles and long bristles on every segment. These sea worms have a head with eyes, short tentacles and a mouth with jaws or teeth.

V This is a sea worm. Each of its body segments has many bristles and two paddles.

paddle with bristles

head

Some sea worms, called tube worms, keep their bodies hidden away inside a tube. The segments closest to a tube worm's head have quite big paddles and long bristles. Further along its body, its paddles and bristles are much smaller and shorter. Many tube worms have eyes, but they do not have teeth or jaws. They have many long tentacles on their heads. A tube worm can stretch its tentacles out of its tube and pull them back inside.

tentacles on top of its head

tube

How do you say it?

microscope: *my-kro-scope*

This tube worm is another kind of sea worm.

7

Special features of annelid bodies

Segments

All annelids have a body made up of many segments. Earthworms and sea worms grow longer by growing extra segments as they get older. The new body segments are added at the tail end of the body. This means that the youngest segments are closest to the tail end and the oldest segments are closest to the head.

Paddles

Sea worms have paddles. These paddles are flaps of skin on the sides of their body segments. The paddles help the worms crawl, swim or grip the sides of their tubes or burrows. They can be different sizes and shapes. Many sea worms have two big paddles on each segment. These form a row down each side of the worm's body. Sea worms that live in tubes often have small paddles on their segments.

Did you know ❓

Most annelids can move backwards if they need to, but leeches can only move forwards.

ⓥ **This sea worm has big paddles on the sides of each segment.**

Bristles

Leeches do not have bristles, but all other annelids do. These bristles are found on the body segments of sea worms and earthworms. Sea worms have many bristles. Their bristles help them crawl or swim. Some sea worms, such as bristle worms, have very long bristles, but others, such as tube worms, have short bristles. These bristles help them grip the walls of their tubes. Most earthworms have tiny bristles. They can move their bristles in and out of their skin. When an earthworm sticks its bristles out, it can grip the ground better so that it does not slip as it wriggles around.

This sea worm has small paddles and many long, white bristles on each body segment.

Suckers

Leeches are the only annelids that have suckers. Suckers let a leech move about in a different way to other annelids. When a leech moves, it grips onto something with the big sucker under the tail end of its body. Then it stretches forward to grip something with its front sucker. Next it swings forward to make a loop with its body and grips again with its big tail end sucker.

small sucker under the head

big sucker under the tail end

These leeches live in the sea. Leeches are the only annelids that have suckers.

Leeches crawl about on their suckers.

The life cycles of annelids

Annelids have three main kinds of life cycle. Two of these life cycles produce different kinds of baby annelids. The other life cycle does not produce any babies. Instead, an adult annelid produces new adults from its body.

Life cycles with babies

Most annelids can make baby annelids by sexual reproduction. In sexual reproduction, an adult female provides eggs and an adult male provides **sperm**. When an egg and a sperm join, a new baby annelid begins to grow.

Annelids have two different kinds of life cycles where baby annelids are made. In one kind of life cycle, the baby annelids first grow into **larvae** that swim about in water. The larvae look very different to their parents. Later, the larvae change to look like tiny adults. The second kind of life cycle does not produce swimming larvae. When these baby annelids hatch, they already look like their parents.

A life cycle with swimming larvae

Most sea worms **reproduce** by making baby annelids called larvae.

To reproduce, an adult female provides eggs and an adult male provides sperm. Most sea worms release their sperm and eggs into the sea at about the same time and place.

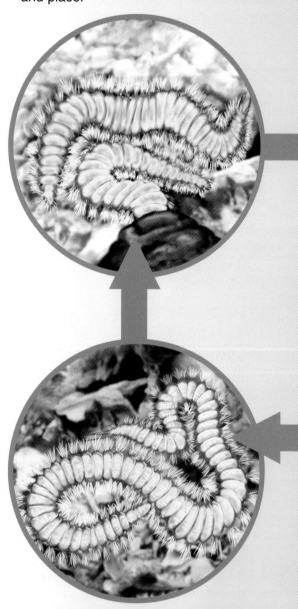

Most of these young annelids become adults when they are between one month and two years old. The adult annelids are then ready to reproduce.

When an egg and sperm join, a larva begins to grow. This larva is so tiny that a microscope is needed to see it. The larva swims about in the sea or near the **sea floor**. It looks very different to adult annelids.

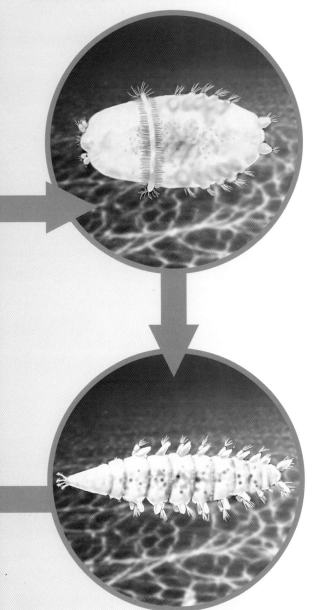

⋀ **This sea worm's life cycle includes a stage with swimming larvae.**

Fascinating fact

Some annelids only live for a few weeks, but others live for a few years or more. Scientists have even found one kind of tube worm that could be 250 years old! This tube worm was recently found growing on the sea floor near North America.

After a while, the larva's body starts to change to look like a tiny adult. This is when the baby annelid starts to grow its body segments. This young annelid is still less than 1 millimetre long. The larvae of some annelids change after a few days, but others remain as larvae for weeks or months.

How do you say it?

larva:	*lar*-va
larvae:	*lar*-vee
reproduce:	*ree-pro-duse*

A life cycle without swimming larvae

Leeches, most earthworms and some sea worms have a different kind of life cycle. These annelids do not produce larvae that swim about in water. Instead, their eggs hatch into baby annelids that already look like the adults.

Leeches and earthworms produce both eggs and sperm inside their bodies. This means that a leech or an earthworm is both a female and a male at the same time. Most of these annelids cannot join their own eggs and sperm together to make their babies. Instead, they need to find another adult annelid to mate with. The two adults swap sperm. Each adult's sperm is placed inside their partner's body, near their eggs.

▽ This earthworm has a saddle on some of its segments. This means that it is mature and ready to mate with another earthworm. The saddle is almost always found closer to an earthworm's head than its tail end.

head

saddle

The eggs and sperm join inside the parent's body and baby annelids begin to grow inside the eggs. These eggs are placed in a small cocoon where they will be protected.

The cocoon is made by a special structure called a saddle. The saddle appears on some of the adult annelid's segments when it is mature. Then, when the adult mates, its saddle makes the cocoon around its body. Up to 20 eggs are placed inside the cocoon. Then the adult wriggles out of the cocoon and closes the ends.

cocoon

cocoon

It can take weeks or months for the eggs to hatch. When they do hatch, the young annelids look like tiny adults. Most of these annelids grow into adults when they are between two weeks and one year old.

A life cycle without babies

Some sea worms and many earthworms that live in water have another kind of life cycle that does not make baby annelids. These annelids do not need to find a partner before they reproduce.

Some of these annelids reproduce by splitting their bodies into two or more parts. Each part then grows back the missing bits to make new adults. Other annelids grow new adults from a special segment near the tail end of their body. Sometimes many worms can be found joined together in a chain.

▼ **This sea worm is reproducing a chain of new sea worms from a special segment near its tail end.**

parent

This is the newest sea worm. It is growing from the tail end of the adult.

This new sea worm is almost ready to break away from the chain.

Where annelids live

Annelids live in many places all around the world. Some annelids live in freshwater ponds, lakes, dams, streams or rivers. Some live on land and some live in salty seas.

Fresh water

Most leeches and some kinds of earthworms and sea worms live in fresh water. Most of these annelids live in shallow water that is either still or slow-flowing, but some live in fast-flowing streams. Many freshwater leeches live in places where there are stones, plants or logs under the water that they can grip onto with their suckers. Some leeches live on soft surfaces such as mud and others just swim about in the water.

Most freshwater earthworms and sea worms burrow into soft mud along the banks or on the bottoms of pools, lakes and streams. Some of these annelids build little mud tubes to live in. Some swim or crawl about over the bottom or on plants in the water. Others swim near the bottom.

Fascinating fact

Scientists can tell if a stream is polluted by looking at the kinds of invertebrates that live there. Most animals cannot live in badly polluted water, but some kinds of earthworms can live in these places. If these worms are the only living things found in a stream, then the stream is probably very polluted.

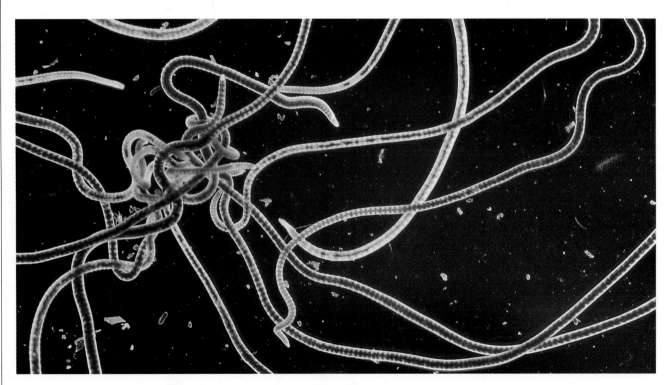

▼ **Some annelids, such as these earthworms, live in fresh water.**

Land

Many earthworms and some leeches and sea worms live on land. Most of these annelids live in moist **habitats**. For example, most earthworms are found in moist soils that are not too dry or sandy. They do not like their surroundings to be too hot, cold or dry. To get away from the summer heat or the winter cold, they may burrow deeper under the ground. Burrowing deeper under the ground also helps them find moist soil. They often come up near the surface of the ground when it rains, or at night when it is cool and damp.

In parts of South-East Asia, Australia and South America, there are leeches that live on land. They are common in wet rainforests where they live on or near the ground. They also live in drier forests, but they stay close to gullies and other damp places. In dry weather, some of these leeches burrow down into the soil where they can survive for many months.

> **Land leeches crawl over the ground and up onto plants.**

Did you know ❓

When an earthworm burrows, it pushes its head into the soil and swallows the soil as it crawls forward. Some of the soil is used as food and some is mixed with a slime called **mucus** inside its body. This muddy mixture is then rubbed onto the walls of the burrow to make the walls stronger.

How do you say it?

mucus: *mew-cus*

The sea

Most sea worms and some leeches live in the sea. Some of these annelids swim near the surface of the water, but most live on the sea floor. They can be found from the deepest parts of the sea through to the seashore.

Ⓐ **Some sea worms live on the sea floor.**

On and under the sea floor

Some sea worms, such as scale worms, live on the sea floor where they hunt other invertebrates to eat. These worms are hard to see because they crawl about under the shelter of rocks, shells and **coral** or under seaweeds that grow on the sea floor.

Many sea worms live in burrows. These burrows are made in places where the sea floor is made of sand or mud. Some sea worms make short burrows to lie in and some make long burrows with many side branches to crawl through. Some sea worms live in burrows under the seashore, but people rarely see them. For example, beach worms and mud worms live under sandy beaches and mudflats, but stay hidden in their burrows.

In tubes on the sea floor

Many sea worms build tubes to live in. Some of these tube worms make hard tubes that they stick to rocks, corals or other hard surfaces. Others make soft tubes. Sometimes these soft tubes are made out of mucus, which is a slime made by the worm's body. The mucous tubes are often made stronger by mixing in sand or bits of shell.

Some tube worms leave their tubes to find food, but others never leave their tubes. The fan worms and Christmas tree worms are tube worms that stay in their tubes. These worms have a beautiful crown of tentacles on their heads. When they are feeding, their tentacles wave about in the water. If they are disturbed, they move back inside their tubes.

These tube worms crawl about over the sea floor and take their tubes with them.

These colourful head tentacles belong to Christmas tree worms. Christmas tree worms live inside hard lumps of coral.

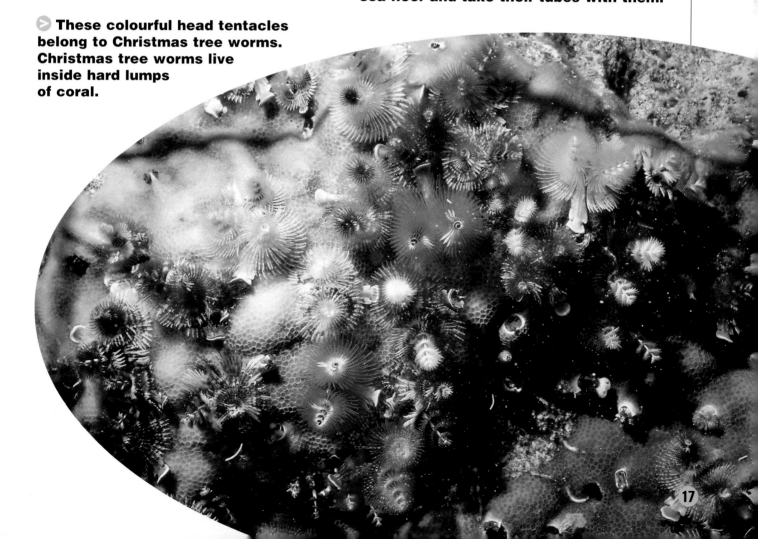

How annelids sense the world

Annelids use a variety of senses to find out about their surroundings. Annelids have the sense of touch, they can see light and dark and they can taste or smell chemicals in the air or water. These senses help annelids escape from danger and find the right food and shelter.

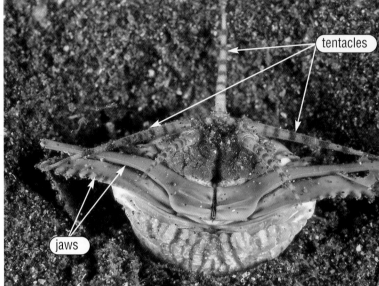

tentacles

jaws

▲ This sea worm is in a burrow in the sand. It uses its tentacles to touch, smell and taste things.

Touch

Annelids can feel things that they touch with their skin. Leeches and some sea worms can also feel when something moves nearby. For example, sea worms and leeches that live in water can feel the tiny ripples that animals make as they crawl or swim past. Leeches that live on land can feel tiny movements on the ground when an animal walks past.

Smell and taste

Leeches, earthworms and some sea worms find their food by smelling and tasting it in the air or water. The skin of earthworms and leeches is covered with many tiny bumps that can smell and taste. Some sea worms have special tentacles on their heads that can taste and smell.

◀ This is a fish leech. A hungry leech will stretch out its body and bend from side to side to smell and taste the water with its skin. This helps it find food.

Did you know

Some leeches can sense whether something is warm or cold. This sense is useful for the leeches that need to feed on warm blood from birds or **mammals**.

Light

An annelid's skin can sense if the surroundings are light or dark. Most annelids hide from bright light and will move underground or into darker places if there is too much light. Many only come out to feed at night.

This tube looks empty, but a tube worm is hiding inside.

Some leeches and sea worms do not have eyes, but many have simple eyes on their heads. For example, a leech can have as many as 10 tiny eyes. Their eyes are very good at telling how light or dark things are and some can see the shapes of things nearby.

 Some tube worms sense light and dark with their tentacles. The shade caused by a passing shadow can make these worms hide inside their tubes. If all these tube worms were to hide inside their tubes, none of their red tentacles would be seen. All you would see is their white tubes.

Fascinating fact

Sometimes an earthworm lies in its burrow with its head or the tip of its tail just poking out of the entrance. This lets it find out how light or dark it is outside because the skin on the top of its head and near its tail can sense light better than other parts of its skin.

Did you know ❓

Some burrowing sea worms can sense their body position to tell if they are burrowing up or down.

What annelids eat

Annelids eat many different foods. Some annelids eat tiny plants and animals and others eat larger ones. Some annelids eat plants and animals that have rotted away into tiny pieces. Some annelids eat blood.

Filter feeders

Many of the sea worms that burrow or live in tubes are filter feeders. These annelids stay in the one place and gather food from the water around them. To do this, they have special food-gathering tentacles on their heads. These tentacles wave about in the water and strain or filter tiny plants and animals out of the water.

▷ **These tube worms are filter feeders. Their feathery tentacles trap tiny bits of food floating in the water. This food is collected and sent to the worm's mouth.**

Fascinating fact

Some kinds of leeches are predators when they are young, but grow into blood-sucking parasites when they are adults.

Predators

Some annelids are **predators** that eat small invertebrates such as snails, insects and other annelids. Some sea worms have strong jaws to grab their **prey** and some even inject their prey with poison when they bite them. Some earthworms and leeches are predators too. They usually catch tiny invertebrates and swallow them whole.

Detritus eaters

Most earthworms and sea worms eat **detritus**. Detritus is made up of plants and animals that are rotting away. When plants and animals rot away, they break up into smaller and smaller pieces and mix in with the soil, sand or mud. Some annelids eat this food as they burrow in the soil, sand or mud. Others send out long tentacles over the sea floor to pick up bits of food from the surface.

▷ **These earthworms burrow through the ground and eat detritus in the soil.**

▷ **This tube worm's body is buried in the sea floor, but it stretches its long, white tentacles out to search for food. Tiny bits of food stick to the tentacles and are passed along to its mouth.**

Parasites

Most leeches are parasites. A parasite attaches onto or inside a larger living thing, then uses parts of it for food. Leeches attach themselves to a larger animal, then feed on its blood. Some leeches only feed on the blood of invertebrates such as snails and earthworms. Others feed on **vertebrates** such as fishes, turtles, birds, dolphins and people. When a leech finds the right kind of animal, it uses the sucker around its mouth to grip the animal's skin. Then it stabs or cuts the skin and sucks up the blood until it is full.

How do you say it?

detritus: *de-**tri**-tus*
prey: *pray*

Fascinating fact

When a leech sucks an animal's blood, it puts two special chemicals into the wound it makes. One chemical makes the animal's skin numb so that it does not feel anything. The other chemical keeps the wound bleeding so the leech can suck up the blood.

How annelids defend themselves

Annelids need to defend themselves from many different kinds of predators. Annelids that live in the sea or in fresh water are eaten by animals such as fishes, **crustaceans**, other annelids and wading birds. On land, annelids are eaten by animals such as slugs, insects and birds. Annelids protect their soft bodies from these predators in a number of ways.

Bristles

Most annelids have bristles sticking out of their skin. These bristles help annelids defend themselves in different ways. Some annelids, such as earthworms and tube worms, have short bristles to help grip the sides of their burrows or tubes. These bristles are like little anchors and make it harder for predators to pull the worm out of its burrow or tube. Other annelids, such as some sea worms, have long spiky bristles to protect them. One kind of sea worm, called a fire worm, has bristles that are like tiny glass needles with poison inside. These bristles can prick a predator's skin and inject a very painful poison.

Did you know

Some giant earthworms squirt a liquid several centimetres out of their bodies when they are disturbed. Some earthworms produce a horrible-smelling liquid when they are touched. The liquid comes out of pores in their skin and makes the worm less attractive to a hungry predator.

 Fire worms have many sharp bristles that can inject poison.

Hiding

Most annelids hide from danger. Some annelids hide under rocks or plants. Others hide in burrows or specially built tubes. Some tube worms stay inside their tube all the time and only stretch part of their body out of the tube to feed. Many of these tube worms can quickly duck back down into their tubes if there is danger nearby. Some kinds of tube worms can even close the top of their tube with a special part of their body that makes a lid.

Ⓐ **Tube worms are good at hiding. This one has hidden its tube in the sea floor. It can only be seen when it has its tentacles out to feed.**

Ⓐ **This sea worm is called a scale worm. Scale worms often hide under rocks and have big scales or plates on their backs to protect their soft bodies.**

Fascinating fact

Fan worms push their feathery tentacles up out of their tubes to feed, but they quickly hide again if a predator comes near. They even hide if a shadow passes over them because the shadow could be made by a predator coming to eat the fan worm.

Growing back missing body parts

Sometimes a predator will attack an annelid and bite off part of its body. Many earthworms and sea worms are able to survive this attack and **regenerate** or grow back the part that is missing. Some sea worms are very good at regenerating parts of their bodies. For example, tube worms often regenerate missing tentacles or even a head. If they lose their head, they quickly regenerate a new one within a few days so they can eat again! Earthworms can also regenerate parts of their bodies, but they cannot regenerate a missing head.

How do you say it?

crustaceans: crus-***tay***-shuns
regenerate: ree-***jen***-er-ate

Ⓐ **This sea worm is regenerating the tail end of its body.**

Blood-sucking

Most leeches live in fresh water, but some live on land and some live in the sea. Some leeches are only half a centimetre long, but most are 2 to 6 centimetres long. The biggest leech in the world comes from South America and grows 30 centimetres long.

Most leeches feed on blood that they suck from bigger animals. Blood-sucking leeches do not need to eat very often, but, when they do, they suck up as much blood as they can. Their bodies can stretch a long way to fit in all this food. This makes them much longer and wider after a meal. Their meal lasts them a very long time. Some kinds of leeches can go without food for one-and-a-half years without dying! Other kinds of leeches need to eat more often and can only live for three weeks without eating.

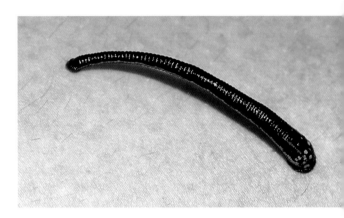

▲ **Before its meal, a leech's body is very thin.**

Fascinating fact

Some blood-sucking leeches have a mouth with a needle-like tube that they push into an animal's skin. Other leeches cut the skin with sharp teeth on tiny jaws. Many leeches have three jaws and leave a Y-shaped cut in the skin. Some Australian leeches only have two jaws and leave a V-shaped cut in the skin.

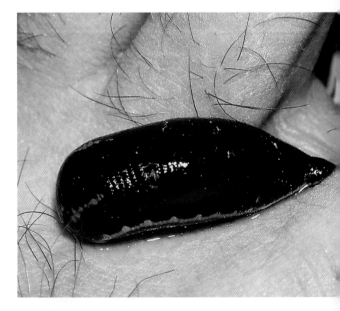

▲ **As it sucks up its meal of blood, the leech's body stretches and swells to become longer and wider.**

leeches

Leeches often prefer to feed on blood from certain kinds of animals. For example, some leeches will only eat the blood of fishes. Other leeches will take blood from a variety of animals, including people. Some leeches that suck human blood are called medicinal leeches because they have been used in medicine.

Hundreds of years ago, people thought that many illnesses were caused by having bad blood. They also thought that leeches could help because they could suck out the bad blood. By the 1930s, people had learned new ways to treat sick people and leeches were hardly ever used. Now, some doctors and hospitals are using leeches again. For example, leeches can be useful when doctors operate to join a finger back onto someone's hand. The leeches remove stale blood from the damaged finger to help it heal.

How do you say it?

medicinal: med-**iss**-in-al

Did you know ❓

A leech takes about half an hour to suck up its meal of blood.

⚠ **This leech is feeding on a shark. The blood it has sucked up into its body can be seen through its skin.**

Earthworms that live on land are the annelids most people know best. There are thousands of different kinds of these earthworms. Most earthworms grow to be about 10 centimetres long, but some kinds are much shorter and some are much longer. The smallest are only 2 millimetres long. One kind of earthworm, called the giant Australian earthworm, grows more than 3 metres long.

> **Most earthworms are long and thin, but this one is short and fat.**

∧ **These giant Australian earthworms live several metres deep underground.**

Earthworms lay one to 20 eggs in a yellowish cocoon. The cocoon is then left in the soil for the eggs to hatch. Earthworms can keep producing eggs and cocoons if they have enough food and if their surroundings are warm and moist. By the time the eggs are ready to hatch, the cocoon has changed to a dark brown colour. Different kinds of earthworms make different sized cocoons. Most earthworms make cocoons that are less than half a centimetre long and wide.

live on land

Earthworms live in burrows in the soil. Sometimes hundreds of earthworms live in the same small area of soil. Some earthworms live close to the surface of the ground and others live deeper. The smallest kinds of worms always live near the surface. The biggest worms burrow near the surface, but, if the soil is deep enough, they will also burrow several metres underground.

Ⓥ **If they have enough food and the right surroundings, hundreds of earthworms can live in a small area.**

Earthworms are very useful animals because they help improve the soil. This, in turn, helps plants grow better. Earthworms improve soil by burrowing through it and by fertilising it. When earthworms burrow in soil, they dig it up and make it looser. This helps air and water get down into the soil where they are used by plant roots. Earthworms also fertilise the soil because they recycle the dead plants and animals they eat into foods that plants can use.

Did you know

Earthworm farming has become very popular. People feed the worms their kitchen scraps and use the fertilisers that the worms make on their gardens.

Where can you see annelids?

The seashore

The easiest place to see sea worms is on the seashore. Crawling sea worms can sometimes be found hiding under rocks on the seashore. Look under these rocks, but remember to leave them the way you found them so that the annelids and other invertebrates that use them for shelter do not die.

Did you know

Some people catch beach worms to use for bait when they go fishing. To catch them, rotting fish heads are dragged over the sand. When the worms smell the fish, they poke their heads up above the sand and are quickly grabbed.

⋀ **Sea worms can sometimes be found hiding under rocks on the seashore.**

Safety tips

- Make sure that an adult is nearby when you explore rocky shores and other seashores. Rocky shores can be dangerous places to explore because the rocks can be slippery and big waves sometimes crash over the rocks. An adult can watch out for these big waves and help you explore safely.

- If you find a sea worm, do not touch it. Some have strong jaws and bite. Some have bristles that sting.

Sea worms can sometimes be found in burrows or tubes. Sea worms that burrow in the sand or mud and sea worms that build soft tubes can sometimes be found in shallow water on sheltered sandy beaches. Sea worms that live in hard rock-like tubes can often be found on rocks close to the edge of the water. Sometimes these worms build their tubes on top of one another. When many tubes are built in the one place, they can look like lumps of rock or coral. The little sea worms that live inside these tubes can only be seen when they are covered by water. When the sea covers them at **high tide**, they stretch their tiny tentacles out to feed.

<ins>The lump around the bottom of this wooden post is made up of hundreds of hard tubes built by tube worms. When the sea comes in at high tide, water covers the tubes and the little worms can feed and build more tubes.</ins>

Gardens

Vegetable gardens and flower gardens are often good habitats for earthworms because they have moist soils and plenty of dead leaves or other food for them. If you dig in the garden, you are likely to find earthworms wriggling about in the soil. The earthworms will be close to the surface if the soil is moist. If the soil is dry, you will need to dig deeper. Sometimes earthworms can be seen crawling about on the surface of the ground. You are more likely to see them doing this at night or after rain.

There are often many earthworms in moist garden soils.

Other places

In bushland habitats, you can look for earthworms under leaves on the ground, under logs and in the soil. It is easiest to find them when the weather is warm and moist. Blood-sucking land leeches can be found in some damp habitats such as rainforests and gullies but, more often, they will use their senses to find you! Leeches that live in fresh water can sometimes be found clinging onto the undersides of rocks or on plants in ponds and streams.

Quiz

1 Is an annelid an invertebrate? Why?

2 Which kind of annelid has suckers?

3 Which kind of annelid has paddles on the sides of its body?

4 What is made by an earthworm's saddle?

5 Name two kinds of annelids that live on land.

6 An annelid uses its skin to sense light. List three other things that an annelid can sense with its skin.

7 What do tube worms have on their heads to help them filter feed?

8 Many annelids can regenerate parts of their bodies. What does this mean?

9 How does the shape of a leech's body change when it has a big meal of blood?

10 Why are gardens a good habitat for earthworms?

Turn to page 32 to check your answers.

Challenge
QUESTIONS

1 How can annelids help scientists find out if a stream is very polluted?

2 If an earthworm lies inside its burrow with only the tip of its tail poking out, how can it tell if it is light or dark outside?

3 When a leech sucks an animal's blood, it puts two special chemicals into the wound. What do these two chemicals do?

4 Why do fan worms hide inside their tubes when a shadow passes over them?

5 Why do some leeches leave a V-shaped cut in the skin but others leave a Y-shaped cut?

The tentacles of a Christmas tree worm are bright and colourful. The rest of the worm's body is hidden away inside the living coral.

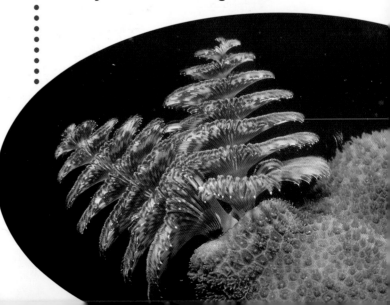

Glossary

corals
Tiny invertebrate animals that live in large groups and make coral reefs. There are soft corals and hard corals. Hard corals have rock-like skeletons that are often used as a habitat by other living things.

crustaceans
(say: *crus-**tay**-shuns*) A group of invertebrate animals that includes crayfishes, crabs, shrimps and prawns.

detritus
(say: *de-**tri**-tus*) Food made up of plants and animals that have rotted away into tiny pieces.

fresh water
Water that is not salty.

habitats
The places where living things, such as plants or animals, live.

high tide
When the sea comes in and covers more of the seashore than at low tide.

larvae
(say: ***lar**-vee*) Baby annelids that live in water. The larvae look very different to the adults. (Larvae = more than one larva.)

mammals
Vertebrate animals that feed their babies with milk and have hair or fur somewhere on their bodies.

microscope
(say: ***my**-kro-scope*) A special magnifying glass used by scientists to see very tiny things.

mucus
(say: *mew-cus*) A slime made in the bodies of many animals.

predators
Animals that hunt other animals to eat.

prey
(say: *pray*) Animals that are eaten by other animals.

regenerate
(say: *ree-**jen**-er-ate*) To grow back missing parts of the body.

reproduce
(say: *ree-pro-duse*) To make more of the same kind of animal or plant.

sea floor
The ground at the bottom of the sea.

segments
Parts of the body that have a similar structure. An annelid's body is made up of many segments. Each segment often looks like a little ring around the annelid's body.

sperm
Special kinds of cells made inside a male animal's body that are used to reproduce. (Cells are tiny building blocks that join together to make a living thing.)

tentacles
Parts of an invertebrate's body that stick out like fingers to sense things or gather food. Tentacles can bend and can often be made longer or shorter.

vertebrates
Animals that have a backbone. They include fishes, frogs, lizards, birds and mammals.

Index

Answers to quiz

1 Yes, because an annelid is an animal that does not have a backbone.
2 A leech.
3 A sea worm.
4 A cocoon to hold its eggs.
5 Earthworms and leeches.
6 Touch, taste and smell.
7 Feathery tentacles.
8 They can grow back missing parts of their bodies.
9 Their bodies stretch and become much longer and wider.
10 Most gardens have moist soils and plenty of food for earthworms.

Answers to challenge questions

1 Scientists look at the kinds of invertebrates living in the stream. If they find certain kinds of earthworms, but no other invertebrates, then the stream is probably very polluted.
2 It senses light through the skin on its body, and the skin near its tail end is very good at sensing light.
3 One chemical makes the skin numb so the animal does not feel the leech. The other chemical keeps the wound bleeding so the leech can suck up the blood.
4 The shadow could be made by a predator coming to eat them.
5 A V-shaped cut is left by leeches with two jaws (each jaw cuts a different side of the V-shape) and a Y-shaped cut is left by leeches with three jaws.